Piaget, Children, and Number

Piaget, Children, and Number

Applying Piaget's Theory to the Teaching of Elementary Number

Constance Kamii and Rheta DeVries

National Association for the Education of Young Children
Washington, D.C.

We would like to thank the teachers and children of Circle Children's Center for their help in illustrating some of the ideas in this book. The teachers in the pictures are: Sargent Aborn, Colleen Blobaum, Robin Reser Burgess, Thomas Gleason, Nancy Fineberg, and Jeanne Klauber. The chidren are: Kenny Bodle, Trenesha Boyd, Jenny Fitzgerald, Jan Gudell, Belinda Hoey, Maggie Houlberg, Deval Jackson, Tanya Krawiez, Bo Lewis, Satimah Lusk, Stephanie Madsen, Lian Mier, Alyssa Migrala, Jennifer Min, Angela Ray, Eve Richards, Matthew Schuler, Jimmy Schuler, and Jesse White.

Cover: Ernest Wesley
Photographs: Constance Kamii and Rheta DeVries

This paper was written as part of a research project supported by the Urban Education Research Program, College of Education, University of Illinois at Chicago Circle, January 1975. The assistance of Hermina Sinclair of the University of Geneva is gratefully acknowledged.

Copyright © 1976, Constance Kamii and Rheta DeVries. All rights reserved.

Second printing, October 1978.

Library of Congress Catalog Card Number: 76-12224

ISBN Number: 0-912674-49-0

Printed in the United States of America.

Contents

An Overview 1

       Notes 4

I. The Nature of Number 5

II. Principles of Teaching 11

       Notes 25

III. Situations Particularly Conducive to the Construction of
     Elementary Number 26

       Notes 46

In Conclusion 49

References 51

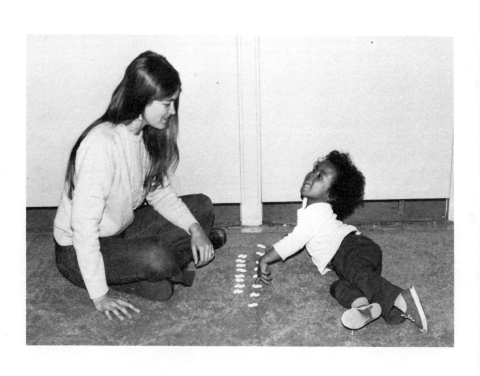

An Overview . . .

When educators hear about Piaget's theory and research on the conservation of elementary number,[1] they inevitably wonder how these can be applied to teaching. Many draw the conclusion that nonconservers should be taught to conserve. Lavatelli (1973), for example, suggested that the teacher remind the child of one-to-one correspondence by making a "bridge" (as shown in Fig. 1) with a pipe cleaner connecting each element of one set with the corresponding element of the other set. In our view, such direct teaching of the conservation task is a misapplication of Piaget's research.[2] When we thus criticize such attempts to apply Piaget's research in the classroom, teachers challenge us by asking, "How, then, do you propose to teach number? Isn't there any way this theory can be applied in the classroom?"

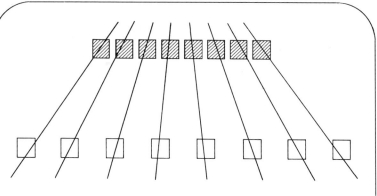

Figure 1. *An attempt to teach children to conserve by making "bridges" to show one-to-one correspondence.*

The purpose of this book is to answer these questions. Piaget's research and theory are indeed useful to the classroom teacher, and they can make a considerable difference in how she or he teaches elementary number. We focus on how the teacher can use the theory in a practical way, by discussing the following three topics:

I. Piaget's theory concerning the nature of number.
II. Principles of teaching number that can be derived from this theory.
III. Situations which the teacher can use to teach according to these principles.

For readers unfamiliar with the conservation of elementary number task, we would first like to review it and Piaget's findings briefly.[3] In this task, the examiner makes a row of 8 objects and asks the child to take out just as many objects from a pile. If the child takes an equivalent number, he or she is asked the conservation question. To ask this question, the examiner rearranges the two equivalent rows as shown in Figure 2 while making sure that the child is watching carefully. The examiner then asks, "Are there just as many here (top row) as here (bottom row), or are there more here (top row) or more here (bottom row)?"[4] The first request to take out just as many is designed to find out whether the child can make a group which is equivalent in number. The second part aims to ascertain whether he or she can conserve the numerical equivalence of the groups when their spatial arrangement is changed so that one row extends beyond the frontiers of the other. Piaget (1941) found the following three levels of reactions to these questions:

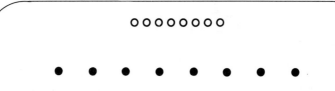

Figure 2. *The arrangement of the objects when the child is asked if the two sets are still equivalent.*

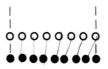

Figure 3. *A typical level I attempt to make an equivalent set.*

frontier   frontier

Level I: **The child cannot make an equivalent set.** The most advanced child at this level typically attempts to make an equivalent set by basing his or her judgment on the spatial frontiers of the rows (Fig. 3). Since he or she cannot even make an equivalent set, there is no point in asking this child the question about conservation.

Level II: **The child can make an equivalent set, but he or she cannot conserve this equivalence.** The Level II child typically uses one-to-one correspondence to make an equivalent set. (That is, he or she puts an object below each one in the examiner's row.) However, when the spatial arrangement of the two rows is changed to that shown in Figure 2, the child typically says that there are more in the bottom row than in the top row. This, despite the fact that the rearrangement was made in front of the child's watchful eyes!

Level III: **The child can make an equivalent set and conserve this equivalence.** When the examiner asks the child how he or she knows that there are just as many in one row as in the other, the child usually gives one of the following explanations:

> Identity ("You didn't take away or add anything. All you did was move the things.")
> Reversibility ("If you move the things back to the way they were before, you'll see that there's the same amount.")
> Compensation ("This line is longer but the things are spread out more.")

Notes

1. This task is described below. *Elementary number* refers to small numbers, such as eight, which the child can conserve without a hierarchical system of inclusions that is required to structure an entire system of number to infinity. (The hierarchical system of inclusions is discussed in more detail in the following section.) Children who can conserve with 8 objects do not necessarily conserve when 15 objects are used. When they conserve with 15 objects, they likewise do not necessarily conserve when 30 objects are used.

Piaget also distinguishes *elementary number* from *perceptual number*. *Perceptual number* refers to numbers up to four or five which are so small that children can make judgments about them based on perception rather than logical reasoning. Piaget warns us to use at least 8 objects in giving the conservation of elementary number task because eight is large enough to preclude judgment by perception.

In this paper we focus on the child's beginning construction of elementary number in a logical sense.

2. This approach teaches children to give correct answers to the specific questions asked in the conservation task, but does not enable them to construct the underlying logico-mathematical structure of number. This statement becomes clearer in the remainder of this paper.

3. The reader already familiar with this research may wish to skip to the section entitled "The Nature of Number."

4. The reader who is skeptical about the child's comprehension of the requests is referred to Kamii (1971), in which both this specific task and Piaget's exploratory method of testing in general are explained in more detail. In the exploratory method, the examiner does everything he or she can to insure the child's comprehension of what is asked. Further details can be found in Piaget (1941).

I. The Nature of Number

In Piaget's theory, number is an example of logico-mathematical knowledge, which is an aspect of knowledge that is different from both physical and social (arbitrary) knowledge. The logico-mathematical nature of number will be discussed first in contrast with physical knowledge and then with social (arbitrary) knowledge.

Piaget makes a fundamental distinction between two types, or poles, of knowledge—physical and logico-mathematical. Physical knowledge is knowledge about objects in external reality. The color and weight of a pencil are examples of physical properties *in* objects which can be known by observation. The knowledge that a pencil will roll when pushed is also an example of observable, physical knowledge about objects in external reality.

When, on the other hand, we are presented with 2 pencils and note that there are 2 pencils, the "twoness" is an example of logico-mathematical knowledge. The pencils are indeed observable, but the "twoness" is a *relationship* created mentally by the individual who puts the 2 objects into a relationship. If the individual did not put the objects into a relationship, each pencil would remain separate and distinct for him or her. The "twoness" is neither *in* one pencil nor *in* the other. It is a relationship, and this relationship is not observable, as it does not have any existence in external reality. Other examples of relationships created by the individual are "longer than," "as heavy as," and "different in color."

Piaget's view about this logico-mathematical nature of number is in sharp contrast with that of the proponents of "modern math." One typical modern math text (Duncan, Capps, Dolciani, Quast, and Zweng 1972) states, for example, that number is "a property of sets in the same way that ideas like color, size, and shape refer to properties of objects" (p. T30). Accordingly, children are presented with sets of 4 pencils, 4 flowers, 4 balloons, and 5 pencils, for example, and are asked to find the sets that have the same "number property." This exercise reflects an assumption that children learn number concepts by abstracting "number properties" from various sets in the same way they abstract color or other physical properties of objects.

In Piaget's theory, the abstraction of color from objects is considered very different in nature from the abstraction of number. The two are so different, in fact, that they are distinguished by different terms. For the abstraction of physical properties from objects, Piaget uses the term *simple* (or *empirical*) abstraction. For the abstraction of number, he uses the term *reflective* abstraction. In simple abstraction, all the child does is focus on a certain physical property of the object and ignore the others. For example, when he abstracts the color of an object, the child simply ignores the other properties such as shape, size, and weight. Reflective abstraction, in contrast, involves the creation of mental relationships between/among objects. Relationships, as we said earlier, do not have an existence in external reality. Numbers for Piaget are thus created by reflective abstraction. The term *constructive* abstraction might thus be easier to understand than *reflective* abstraction—to indicate that this abstraction is a veritable construction by the mind rather than a focus on properties that already exist in objects.

The distinction between the two kinds of abstraction may seem trivial and academic while the child is learning small numbers, say, up to 10. When he goes on to large numbers such as 999 and 1000, however, it becomes clear that it is impossible to learn every number all the way to infinity by abstraction from sets of objects or pictures! Numbers are learned not by abstraction from sets that are already made but by abstraction as the child constructs relationships. Because these relationships are created by the mind, it is possible to think of numbers such as 1,000,002 even if we have never seen 1,000,002 objects in a set.

Number for Piaget is a synthesis of two kinds of relationships the child creates among objects. One of these relationships is ordering, and the other is class inclusion. All teachers of young children have seen the common tendency among children to count objects by skipping some and counting some more than once. When given 8 objects, for example, a child who can recite, "One, two, three, four . . ." correctly up to ten may end up claiming that there are 10 things by "counting" as shown in Figure 4(a). This tendency shows that the child does not feel any logical necessity to put the objects in an order to make sure he does not skip any or count the same one more than once. The only way we can be sure of not overlooking any or counting the same object more than once is by putting them in an order. However, one does not have to put the objects literally in a physical order to put them into an ordered relationship. What is important is that they be ordered mentally as shown in Figure 4(b).

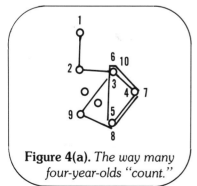

Figure 4(a). *The way many four-year-olds "count."*

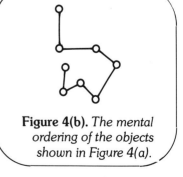

Figure 4(b). *The mental ordering of the objects shown in Figure 4(a).*

If ordering were the only mental action on objects, the objects would not be quantified, since the child could consider one at a time rather than a group of many at the same time. To quantify the objects as a group, he or she has to put them into a relationship of class inclusion as well. This relationship, shown in Figure 5, means that the child mentally includes "one" in "two," "two" in "three," "three" in "four," etc. The child thus may be presented with 8 objects, but he or she will "see" a set of 8 only if he or she puts all 8 of them into a single relationship.

Piaget's theory of the logico-mathematical nature of number is also in contrast with the common assumption that number concepts can be taught by social transmission like arbitrary social knowledge, especially by teaching children how to count. Social knowledge is knowledge which is built neither by observation of physical phenomena (physical knowledge) nor by the creation and coordination of relationships (logico-mathematical knowledge) but by social consensus. Examples are the knowledge that Christmas comes on December 25, and that a chair is called "chair." The characteristic of social knowledge is that it is mostly arbitrary. The fact that Christmas comes on December 25 rather than on December 20 is an example of the arbitrariness of social knowledge. The fact that there is any Christmas at all is another example of the arbitrary nature of social knowledge. There is no physical or logical reason for December 25 to be considered any

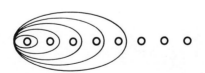

Figure 5. *The structure (relationship) of class inclusion in the child's conception of number.*

different from any other day of the year. The fact that a chair is called "chair" is likewise completely arbitrary. In another language, the same object is called by another name. It follows, therefore, that social knowledge can be taught only by social transmission.

People who believe that "number facts" should be taught by social transmission fail to make the fundamental distinction between logico-mathematical and social knowledge. In logico-mathematical knowledge, nothing is arbitrary. For example, $2+2$ gives the same result in all cultures. In fact, every culture that builds any mathematics at all ends up building exactly the same mathematics, as mathematics is built on the internal consistency of a deductive system in which absolutely nothing is arbitrary. To cite another example of the nonarbitrary nature of logico-mathematical knowledge, in all cultures, there are more animals than cows.

In conclusion, Piaget's view is in contrast with the belief that there is a "world of number" into which each child must be socialized. To be sure, even two-year-olds can see the difference between a pile of 3 blocks and one of 10 blocks. This, however, does not imply that number is "out there" in the physical world, to be abstracted by simple (empirical) abstraction. By the same token, even though there is consensus about the sum of $2+2$, number is not "out there" in the social world, to be learned by social transmission. We must be aware that the arithmetic we can teach directly is limited to certain arbitrary social conventions, such as how to say, "One, two, three, four," and the details of specific computation, such as $5+5$, which can be taught only after the child himself has constructed the logic underlying the addition of two numbers.

If the child cannot create relationships such as "five" in his mind, no amount of verbal counting or drill will enable him or her to construct the notion of "five." While we thus believe with Piaget that number concepts are not directly teachable, we do not draw the pedagogical implication that the only thing the teacher can do is sit back and wait. There are certain things he or she can do to encourage children to construct number concepts. These are discussed in Section II as principles of teaching that we derive from Piaget's theory of number.

II. Principles of Teaching

1. Teach number concepts when they are useful and meaningful to the child.

We have emphasized elsewhere (Kamii and DeVries, in press) the most basic principle of teaching we derive from Piaget's constructivism: Promote the child's autonomy. If children develop by constructing their own knowledge and morality, they must be encouraged to act out of their own choice and conviction. Thus, we would not advocate that young children between the ages of four and six have a time set aside just for math. Rather than doing math because the teacher says to do it, children should be encouraged to think about number when they feel a need and interest. All children between four and six years old seem to be intensely interested in number when its use is at the right level for them. They spontaneously count the presents they receive and the marbles they own, and argue about who has more blocks than they have. They also adore songs and board games involving counting. In Section III of this book, we discuss the kinds of situations in which the teacher can capitalize on this spontaneous interest.

2. Use language that elicits logical quantification and the comparison of groups.

Logical quantification is contrasted here with numerical quantification. An example of using the language of logical quantification is "Bring *just enough* cups for everybody at your table," instead of saying "Bring *six* cups." Although we feel it is good for children to learn how to count, we feel that when the teacher is trying to get children to construct beginning number concepts, it is best to avoid telling them to count. The reason for this belief is that, for preoperational children, the ability to count is one thing, and quantification is quite something else. In the conservation of elementary number task, for example, many Level II children count the two equivalent rows correctly, but still say that there are more in one row! Some children at this point occasionally look embarrassed and confess to having made a mistake. Most Level II children, however, see no contradiction between saying that each row has eight, and saying that the bottom row has more.

Counting thus helps the Level II child only once in a great while to conserve. If it helps him or her, this is because the child's level of cognitive development in logico-mathematical knowledge is already quite high, and he or she is very close to conserving anyway. If, on the other hand, the child's cognitive development is low, no amount of counting can help him or her to conserve. Counting for this child is like saying, "Mark, John, Suzy, Mary, Bobby, . . . Judy," in the right order, and answering the question "How many?" by repeating the last name in the series. Thus counting causes the child to focus on each object and its name in the series, rather than the entire set. This is why we say that counting is one thing, and quantification is quite another thing.

When asked in the conservation task, "Are there as many here as here?" the child compares the two sets in the best way he or she can. If the child does not have the logico-mathematical structure of number (which is built by ordering objects and putting them into a relationship of class inclusion), the best way he or she can quantify the sets is by comparing the spatial frontiers of the two sets. Similarly, when the teacher asks a child to "bring just enough straws . . . " the child will quantify in the best way he or she can. If the child can use counting, he or she will. If the child has not yet constructed his or her notion of number, and therefore cannot see the use of counting, the child will still reason about the numerical issue in some way. He or she may use one-to-one correspondence, or grab a fistful and end with too many or not enough. Still, the child has thought in some way about a quantitative problem.

Thus, we recommend that the teacher avoid asking a child to count because this gives a "trick" or half of the solution, rather than encouraging him or her to figure out what to do. Other examples of language that elicit logical quantification and comparison of sets are:

> Do we have *enough* for everybody to have one?
>
> Do we have *too many* cups?
>
> Do you have *as many* chips (or the *same amount,* or the *same number*) as I have?
>
> Who has *more?*
>
> Bobby has *less* than you do.

3. Encourage children to make sets with movable objects.

When we encourage children to focus on only one set of objects at a time, we limit their possibilities for thinking about quantity. The only type of question we can ask about a single set is of the kind "How many are there?" or "Can you give me eight?" We thus can only end up encouraging children to count the objects. As explained in Principle 2, encouraging children to count is not a good way to help them begin quantifying groups of objects. A better approach is to ask them to compare two sets.

There are two ways of asking children to compare sets: by asking them to *make a judgment* about the equality (or nonequality) of sets that are already made, and by asking them to *make a set* that has the same number as another set. The second approach is far better for two reasons. First, when a child is asked to make a judgment about two ready-made sets, his reason for comparing the sets is that the adult wants an answer to the exercise he or she chose for the child. Comparing ready-made sets is a passive activity in which the child is limited to only three possible responses: The two sets are the same; one has more; or one has less. When the child has to *make* a set, in contrast, as when he or she is asked to bring just enough straws for everyone at the table, the child starts with zero, takes one, one more, one more, etc., *until he or she decides when to stop.* This kind of decision necessary to the physical making of a set has more educational value because there is a greater degree of freedom in starting with zero and knowing exactly when to stop the action of "adding one more." In making sets, children have a chance to order the objects actively and group them.

The value of encouraging children to make sets implies that some commonly used materials are inappropriate for teaching elementary number. Workbook pictures such as Figure 6 and Cuisenaire rods (Kunz 1965) are examples of such unfortunate materials.

Figure 6. *An example of a typical workbook page.*

Bits of styrofoam packing, vegetable trays, and dice are used in this game. Starting with about 20 to 25 pieces in the middle, players take turns throwing the dice and putting that number of pieces into their trays. Here, since all the pieces have been taken, it is time to count to see who got more.

When this child had the last turn, she threw a "6," but there were only 4 pieces left. So, looking dubiously at the teacher, she took two additional pieces from the can of materials. For her, having to stop at 4 was not fair.

Exercises such as the one shown in Figure 6 are undesirable because they preclude any possibility for the child to move the objects and make a set. Moreover, such an exercise easily elicits the kind of reasoning that yields the right answer for the wrong reason. For example, when asked how they knew the right answer to Figure 6, many children explain, "You draw lines like this, and then put an X on the set that has one left over." Such children may or may not have the slightest idea about which set has *more*. If they do, this is usually because they already have the ability to quantify objects. If they don't, the exercise is useless because children do not learn to make quantitative judgments by drawing lines on paper.

Cuisenaire's approach to teaching number with rods reflects the common confusion between discrete and continuous quantities. For Cuisenaire, the 1-cm rod stands for "one," the 5-cm rod stands for "five," and the 10-cm rod stands for "ten." For Piaget and most young children, however, each of these rods can only be "one," since it is a single, discrete object. Number involves the quantification of discrete objects, and therefore cannot be taught through length, which is a continuous quantity.[1] Giving such a ready-made "two," "three," "four," etc., to children is worse than giving them ready-made sets such as Figure 6. At least these are sets of discrete objects.

Montessori (1912), Stern (Stern and Stern 1971), and many others, too, made seriated rods proportioned according to the same principle as the Cuisenaire rods. This principle is to make the second rod twice as long as the first one, the third one three times as long as the first, etc. Advocates of teaching number with such rods believe that by ordering the rods, children learn about the number series, including the idea that "one" is included in "two," "two" is included in "three," etc. Piaget's research shows that, in reality, when the child arranges the rods from the longest to the shortest, or vice versa, all he or she learns is the crutch of how to use the stairstep shape to judge whether or not his arrangement is correct. This shape is an observable spatial configuration, which the child can use as a source of external feedback. In logico-mathematical knowledge, feedback can come only from the internal consistency of a logical system constructed by the child. This system, as can be illustrated by the system of class inclusions, is not observable. Ability to arrange objects by trial-and-error based on feedback from the configuration is thus not the same thing as the development of logic.

4. Get children to verify an answer among themselves.

As stated above, arithmetic does not have to be transmitted from one generation to the next like social (arbitrary) knowledge, since all children in all cultures construct the same arithmetic if they construct any at all. Therefore, if children are encouraged to think about numerical quantity, they are bound to construct notions which will eventually lead them to right answers. When a child brings "just enough straws . . .," therefore, the best thing for the teacher to do is to refrain from giving direct feedback and encourage the same child or other children to check the answer. When children are confronted with an idea that conflicts with their own, they are motivated to think about the problem again, and either revise their idea or argue for it. Therefore, an important principle of teaching is to bring disagreements to children's attention by asking for many opinions or by mentioning casually to a child that someone else has a different idea.

When we teach number by being the only source of feedback, we unintentionally teach many other things, such as how to read the teacher's face for signs of approval. Such teaching amounts to education by conformity to the person in authority. This is not the way children can develop confidence about their own ability to figure things out. Piaget (1948) vigorously opposes this kind of teaching and insists that the emotional block many students develop about math is completely avoidable.

5. Figure out how children are thinking.

If children make an error, it is usually because they are using their intelligence by reasoning in their own way. Since every error is a reflection of the child's thinking, the teacher's task is not to correct the answer, but to figure out why the child made the error. Based on this insight, the teacher can sometimes correct the process of reasoning, which is not the same thing as correcting the answer. For example, if the child brings one less than "just enough straws . . . ," the reason may be that he or she did not count himself or herself. Preoperational children often have difficulty considering themselves as both the counter and the counted. When they count the others, therefore, they frequently do not count themselves. When they distribute straws and find they are one short, a casual question such as, "Did you count yourself when you counted the children?" may be helpful.

Just as there are many ways to get the wrong answer, there are many ways to get the right answer, and not all of them are based on logical reasoning. One illustration of this is a study by Piaget (1941, Chapter 8) of how children divide 18 counters between 2 people. He found three different ways (levels) of getting the right answer, only the last one of which is based on logical reasoning. Below are the three levels:

Level I: *A global (intuitive) method*
The child divides the counters in a rough, global way and may by accident give 9 to each person. This is an example of getting the right answer for the wrong reason. After thus dividing the counters, however, the child may end up saying that there are more in one bunch, especially if the spatial configuration is changed as shown in Figure 7(a).

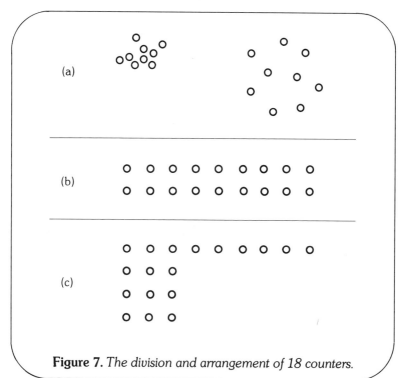

Figure 7. *The division and arrangement of 18 counters.*

Level II: A perceptual (spatial) method

>The child spatially puts the counters in one-to-one correspondence as shown in Figure 7(b). After thus dividing the counters, however, he or she usually ends up saying that there are more in one set if one of the arrangements is changed as shown in Figure 7(c).

Level III: A logical method

>The child gives one (or more) to each person alternately until all 18 of the counters are exhausted. The spatial correspondence is unnecessary and irrelevant when the child's logic has thus developed.

This is a commercial toy called "Don't Spill the Beans" (1967). The players take turns placing a bean on the pot, trying not to make it tip over. The player who spills the beans by placing the one that causes the pot to tip over has to take all those spilled. The winner is the player who gets rid of all his or her beans first.

The game begins by dividing nearly 200 beans among all the players. The way four- and five-year-olds divide the beans shows a "logic" which is not well structured. This five-year-old alternately gave 1 to each player, then 2 to each, switching the pattern randomly or when her hand happened to be empty. (She usually grabbed a handful out of the pot to divide, and then grabbed another handful when her hand was empty.) The teacher observed that the child sometimes skipped a player and sometimes gave to one twice in a row.

When the division was finished, the teacher asked, "Are you sure I have just as many beans as you do?" The child replied, "No." "How can we be sure?" the teacher asked, and the child suggested, "If we count them," and enthusiastically began by saying, "1, 2, 3 . . ." as she dropped one bean at a time into her dish. The counting sometimes went like "45, 46, 48, 49 . . . What comes after 49?" and "56, 57, 59 . . . What comes after 59?"

Without saying anything, the teacher made a straight line with her beans in order to count them. The child was not at all impressed with the teacher's method but helped her count them and decided that the teacher had 5 more than she did. When asked, "What can we do so we have the same?", the solution proposed by the child was "Put these 5 in the pot."

By observing the child's behavior, the alert teacher can infer whether the child is approaching a problem in a global, perceptual, or logical way. On the basis of this kind of continual study, the teacher can give problems to each child at a level that is truly appropriate for him or her.

6. **Encourage children in a general way to put all kinds of objects, events, and actions into relationships.**

Number concepts are not isolated ideas that develop independently of all the other relationships the child creates and coordinates. The child's construction of the class-inclusion structure may serve to illustrate this point. Class inclusion is a cognitive structure characterized by mobility of thought which enables us to think about "five," for example, in relation to "one," "two," "three," "four," "six," "seven," etc. (Fig. 5, p. 8). The ability to coordinate the elements in a hierarchical system requires thought which is both mobile and well-structured.[2]

Mobility of thought develops as the child puts things and events into relationships and structures them in his or her daily life. For example, bedtime may be the same every night except Fridays and Saturdays. There may be items of clothing (such as underwear) which are for every day of the year, and others which are only for parties or Sunday school. In playing a game like bowling, too, the child may notice that the target is missed by rolling the ball too far to the right sometimes, and too far to the left at other times. These are examples of the countless situations in daily life where the child can put elements of his or her reality into hierarchical relationships. The hierarchical relationships develop by the coordination of more limited local relationships, such as "I missed the target again." The child who has created and coordinated many relationships in his daily life is probably the one who is highly "ready" to construct number concepts—even by the worst teaching method.[3]

These are three-year-olds. The teacher, therefore, did not ask them to count all 10 of the pins at the beginning, nor did she make any attempt to compare children's performance or keep score. Each player took a turn and the teacher waited until there were only three or four pins left standing up to ask, "How many do you still have to knock down?" Each child rolled the ball as many times as necessary to knock down all the pins. At age three, a game is more of a game if it does not have a winner or a loser!

In this balance game a ball rests in a cup, and a cardboard disc is balanced on top. The disc is marked off into numbered segments. Players take turns throwing a die and then choose an object to place in that segment on the disc. (Objects vary in size and weight from a small bit of cloth to a small block.) Players try to keep from making the disc fall off. Using two dice makes addition necessary. Everyone holds their breath when the teacher takes a turn!

While the primary value of this game involves physical knowledge, number does play a small part. We include this example to suggest an interesting way for children to practice reading numerals and counting.

We now turn from general principles of teaching to specific classroom situations which lend themselves particularly well to teaching number concepts in a Piagetian sense. We would once again like to emphasize, however, that number concepts cannot be taught by the teacher, since the only way they can be learned is by the child's own construction. If a child has difficulty in constructing beginning number concepts in response to the teacher's use of Principles 1 to 5, it is more fruitful to encourage him or her to build mental connections in a general way by putting people, objects, and events into relationships than to try to teach the child number concepts directly (Principle 6).

Notes

1. A continuous quantity such as a length can be quantified only by introducing an arbitrary unit which is not given in the object. The logical structure of serial inclusion is the same in number and length. In number, however, the unit is given by the discrete object. Continuous quantities may, therefore, be of use to later work in mathematics, but they are completely inappropriate for the teaching of beginning elementary number.

2. The reader is referred to Inhelder and Piaget (1959, Chapters 1, 2, and 4) for a fuller understanding of the significance of this statement.

3. This emphasis on the child's creating and coordinating relationships implies that number should be taught in the context of a total curriculum which is not limited to the teaching of number. The reader interested in a broader curriculum is referred to Kamii and DeVries (in press).

III. Situations Particularly Conducive to the Construction of Elementary Number

The foregoing principles of teaching can be put to work particularly well in situations such as the distribution of materials, games, art, and music.[1] These are discussed below. In these situations, we hope the teacher will remember that elementary number is best taught casually and, for the most part, as an incidental aspect of activities. We hope he or she will not misunderstand the way in which we mean the above principles of teaching to be used. Although we discuss specific kinds of situations with specific examples, we do not mean to imply that number is the primary objective of the activity or that the teacher must always, without fail, teach number in these situations. What we hope is that the teacher will become alert to the possibility of teaching number in a variety of situations and will then decide whether to focus on number or follow children's interests in another direction. We must also caution that any of these situations can be used with too much teacher authority or too much emphasis on getting the right answer.

Quantification as a Part of Everyday Life in the Classroom

Some children come to kindergarten already able to conserve elementary number and to count and compare sets. These children built all these abilities without any lesson, kit, special materials, or workbook. They constructed their logico-mathematical framework in the course of adapting every day to reality by structuring relationships among objects, people, and events, and making sense out of their experience. A part of this construction is quantification, as when children make sure that others do not get more candy than themselves. Quantification is thus an inevitable part of living, and life in the classroom regularly entails quantification. For example, materials must be distributed, things must be divided fairly among children, milk and lunch money must be collected, records must be kept, and we must make sure not to lose the checkers. These responsibilities are usually carried out by the teacher because the tasks are too difficult for children four to six years old. With some planning, however, the teacher can turn these jobs over to the children, at least in part, and create situations in which number can be taught in a natural, meaningful way. Specific situations elaborated below are the distribution of materials, the division of things, the collection of things, keeping records, cleanup, transition times, and voting.

The Distribution of Materials

In Principles 2 and 3, we gave the example of asking children to bring "just enough straws so everybody will have one." This example also illustrates how to make educational use of situations where there is a need for distribution. Since it is not possible for most young children to think about the total number of children in the class, the teacher will probably need to divide it into subgroups sized according to the number the children can handle. As mentioned in Principle 5, children who can use counting in other kinds of situations are often unsuccessful in this task because they "forget" to count themselves. For young children, it is difficult to be both the counter and the counted. Whether it is himself or herself or someone else who gets left out, the counter finds out clearly when he or she has not brought enough. The teacher's task then is to use this social feedback in a positive way as discussed in Principle 4.

The Division of Things

Snacks such as raisins and roasted pumpkin seeds present the problem of dividing things equally among children. In a distribution problem the child knows ahead of time how many things are to be given to each child. In a division problem, on the other hand, he or she does not know the number that each will get. Rather than taking a small subset from a larger set, the child must take a large set and divide it into equal subsets.

Again, this task is too hard if one child has to divide things among the entire class. The teacher can, however, plan to give a certain number to each pair of children (or to each group of three, four, etc.) to divide in a fair way. If a child protests that someone else got more, the teacher should follow Principle 4 and encourage children to continue to discuss among themselves how to get a fair division.

The Collection of Things

The collection of such things as parental permission slips prior to a field trip provides a natural opportunity for teaching the additive composition of number. At group time, the teacher may present the question of whether or not to call the bus company to make the final arrangement. She or he may raise such questions as the following:

> Do we have all the slips we need?

The child in charge of snack counts how many people are in the room (both children and adults), gets out enough cups and napkins for everybody, and puts the same number of marshmallows on each plate. She then gets a plateful of crackers to pass around. There is no shortage of supervisors! The teacher, therefore, refrains from intervening.

How many do we still need to collect?

How many children got a slip yesterday?

How many brought the slip back today?

Who was absent yesterday? How many were absent?

As usual, if the task can be made more approachable by dividing the class into subgroups, this is by all means worth doing. Small groups also offer the advantage of encouraging more children to take leadership and share responsibility.

When the teacher distributes the slips for the children to take home, he or she can say, "I will write on the blackboard how many slips I gave out so I can remember how many to collect." The names of the children who are absent can also be written on the board. This, by the way, is a good incidental way to teach reading readiness.

The collection of lunch money may be too difficult, but milk money may be manageable. At least children can help count the number who paid in order to know how many milks to distribute.

Keeping Records

The chart shown in Table 1 is an example of the kind of atten-

Name		
Adam		
Bobby		
Cathy		
David		
Evelyn		
Total	5	
Present		
Absent		

Table 1. *An attendance chart.*

dance record some young children can handle. This kind of chart is good for teaching not only addition and subtraction, but also how to organize information in a table.[2] The teacher may also want to refer to the attendance chart to figure out, for example, how many more permission slips to collect. As usual, the teacher should include in each chart no more than the number the particular group of children can handle.

Cleanup

Various opportunities for numerical quantification can be found in the context of cleanup. For example, the teacher can label boxes so that children will know how many objects they should contain. The "Cat and Mouse" game (1964) can be labeled to indicate that it should contain 16 mice, 4 each of red, blue, yellow, and green. This might be done by putting "4" or " : : " beside each of four colored lines. The die may also be represented in some way to encourage children to make sure there is a die in the box.

If the teacher has a general cleanup time, he or she might sometimes suggest that everybody put three things away. Some teachers have a chart showing who is responsible for cleaning up each one of several areas of the room. Each person in charge decides how many helpers are needed and invites these helpers.

Transition Times

We have seen teachers take advantage of transition times to give a number problem to each child before he or she gets his or her coat or lunch or moves into free play. Such problems involve the production of a certain number of discrete actions, such as "Clap your hands five times," or "Tap your nose five times and two more." In this situation, the teacher should work toward establishing the kind of atmosphere that will lead chil-

At the beginning of the year, four- and five-year-olds feel no need to vote for only one name for the goldfish. Some vote twice, others three times! Given many opportunities to vote, however, by midyear they do not have to be reminded that they can vote for only one of the choices. They also propose a vote whenever a group decision has to be made.

dren to observe and evaluate each person's performance. Correction should thus come from other children, rather than the teacher.

Voting

On many occasions, the teacher can suggest that the group decide by majority vote. Voting is useful, for example, when the group cannot agree on what to name the gerbil, or whether to plan biscuits or waffles for the next day's snack. While voting does teach number, its more important function is to place the power of decision making in the hands of the children, rather than resorting to adult authority.

Group Games

Many group games provide an excellent context for thinking about numbers. Some examples are given below of board games, cards and dominoes, aiming games, racing and chasing games, hiding games, and rituals.[3] In these activities, it is again important to emphasize that the teaching of number is incidental to more basic objectives pertaining to children's moral and socioemotional development (Kamii and DeVries, in press).

Board Games

"Hi-Ho! Cherry-O" (1972), "Candyland" (1955), "Cat and Mouse" (1964), and "Chutes and Ladders" (1956) are examples of commercially available board games which are good for the child's construction of number. In these games, a number is drawn from a pile of cards, spun with a spinner on a number wheel, or thrown with a die. This number tells the player how

These children are playing "Hi-Ho! Cherry-O" (1972), a counting game in which players begin by hanging ten plastic cherries on a tree. (The teacher found that some children have difficulty making the cherries stay on, and may find it easier if the trees are placed flat on the board as shown in this picture. Children can also see the number of elements better this way.) In turn, each player spins the spinner to see how many cherries to put in the basket. The winner is the one who gets all the cherries off the tree first.

There is no problem of interpretation when the spinner points to 1, 2, 3, or 4 cherries. However, when it points to the segment showing 13 cherries spilling out of the basket, these three-year-olds interpret the picture simply as "a lot." "A lot" to some means to take 4 while to others it means to take all 10. (The rules of the game state that the picture of the basket of spilled cherries means to put all cherries in the basket back *on* the tree. However, to these children this picture means to take a lot of cherries *off* their tree!)

When there are three players, it is hard for young children to keep the order of turns straight. A child may conclude that he is next because the player next to him has just finished—but this depends on the direction the turn is moving!

This is a homemade version of a game called "Tug O'War" (*56 Games,* 1956). Two players each choose an end of the "rope," and a single marker is placed in the middle white circle. Players take turns throwing a die and moving the marker the number of spaces thrown toward his or her end. For example, the teacher here threw a "4" and moved the marker four spaces toward his end. The winner is the one who moves the marker all the way to his or her end of the rope. Since there is only one marker, this game is not a race, but players have the possibility of finding out that a big gain for one player is a big loss for the other.

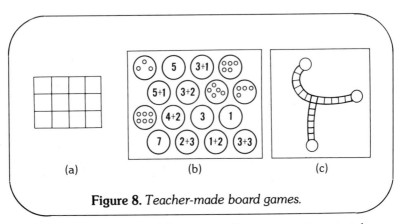

Figure 8. *Teacher-made board games.*

many cherries to take off the tree, or how many steps to take toward the goal. Players have the opportunity to compare their progress with that of other players, and to figure out what number they want to get next. The teacher can simplify the die by including on it numbers only up to three or four. Or, it can be made more difficult by using two or three dice to make addition necessary.

Board games can also be made by the teacher. The one shown in Figure 8(a) is simply a card squared into twelve sections which is used with a die and paper clips or small blocks. The object of the game may be just to fill one's card first, or fill it up and then empty it by taking off the number of objects thrown with the die. The game shown in Figure 8(b) is made from a styrofoam fruit tray discarded from a grocery store. Children can play this like bingo, taking turns rolling a die (or dice) and seeing who can fill up his tray first with the right number of poker chips.

Until children have had considerable experience with board games, the teacher needs to introduce the rules. Afterwards, however, he or she can make a board such as shown in Figure 8(c), and ask children if they would like to make up some rules for a game with it. Eventually, children enjoy drawing their own boards and making up their own variations of board games.

Cards and Dominoes

In card games such as "Old Maid," players match pictures, discard completed pairs, and try not to be left holding the odd card. This game can also be played with regular cards or teacher-made cards with numerals or groups of dots or pictures. In "Concentration," cards are arranged in neat rows, face down, and players try to match pairs by turning one card over and trying to remember the position of its mate. In all these games, children can be encouraged to see how many pairs they have at the end, or who has the most pairs. It may also be possible to use two sets of dominoes instead of cards (with larger numbers omitted if they are too difficult). Dominoes can also be used in a modified version of the adult game, simply by matching halves, or by having players announce the sum of the two ends after making a match.

The reader may be wondering why we reject ready-made sets such as shown in Figure 6, but recommend them in cards and dominoes. We distinguish these two kinds of ready-made sets for three reasons. First, when the child decides, for example, whether a card having six hearts matches one having seven, he or she must truly compare the two. The decision cannot be made by using the crutch of drawing lines as in Figure 6 (p. 00). Second, children have the opportunity to give and receive the kind of social feedback from peers which is more desirable than adult correction (Principle 4). Third, the child is intrinsically motivated to think about numbers and numerals when he or she chooses to play these games. Such motivation is often absent when workbooks or math time is imposed by the teacher.

Cards and dominoes are also movable objects which can be quantified. The dealer may thus have the task of giving an equal number to all the players, putting the rest in the middle, or the task of distributing all the cards to all the players, in which case some children may get one more than the others. This offers the teacher an opportunity to say, "I got 13. How many did you

After dividing the cards equally, each player turns up the top card of his or her pile. The player whose card shows more objects than the other takes both cards. The winner is the player who ends up with more cards. The five-year-old spectators make better "teachers" than those who hold teachers' certificates!

Sometimes, the teacher needs to limit the deck to numbers up to five, thus limiting the deck to a total of 20 cards. A five-year-old who ends up with 10 cards sometimes insists that he has more, even though his opponent has just told him that he, too, has 10 cards! When confronted with an opinion that no player has more, one five-year-old hid one of hers and then insisted that she now had one more when she really had one less! During such arguments, the best thing for the teacher to do is to ask other children what each one of them thinks. If she or he wants to express an opinion, the teacher must be careful to present his or her idea casually as only one of the answers being offered. Posing as *the* authority who knows *the* truth is bad because it undermines children's confidence and initiative to figure out the truth by themselves.

get?" In the course of drawing from a pile, the child may notice that the number in his hand increases when he or she does not make a pair and decreases when a match is found. In games like "Old Maid," the child has the opportunity to notice that when he or she takes a card and gives one away, the number stays the same.

Aiming Games

Aiming games are especially good for counting and making comparisons. Marbles, "Drop the Clothespins," and bowling are examples of such games in which children can be motivated to think about how many objects they succeeded with. The teacher must be careful, however, not to insist on children comparing how many they got. At age four, children are usually interested only in what they themselves did. It is not until age five or six that some begin to take an interest in comparing performances.[4] One more word of caution: When a four- or five-year-old says, "I got seven, and Johnny got eight," he is not necessarily comparing two sets. He may merely be announcing the result of counting.

In "Drop the Clothespins," children climb on a chair or some other object, and try to drop clothespins into a plastic bottle. If each player has 10 pins and a bottle to begin with, he or she can find out how many out of ten were dropped successfully, and how many missed the bottle.

In marbles and "Drop the Clothespins," children usually feel no desire to keep score because they collect their objects and count them at the end of the game. In bowling, however, where children take turns at aiming, they are often interested in counting the number of pins (or blocks or boxes) they knocked down over several turns. The easiest way for them to begin keeping score may be simply by making one mark for each fallen pin.

This unwieldy representation can eventually be structured to
" / / / / /   / / / / /   . "

Racing, Chasing, and Seeking Games

Some games such as "Musical Chairs," "Duck, Duck, Goose," and "Fourteen Stand Still" involve number or number-related activity.

In "Musical Chairs" children can be encouraged to decide how many chairs they need. This game should not be played with the entire class in one group. If children are encouraged to initiate their own games during free play, the number of interested participants is usually small enough. For young children, the classical rules of "Musical Chairs" should be modified so that no one is forced to drop out of the game. Strangely enough, at ages four and five, children generally prefer to play this game with the number of chairs equal to the number of children. They should be encouraged to discuss the rules and decide among themselves whether they want the number of chairs equal to the number of players, or whether they want to use one or two less (without putting anyone out who fails to get a chair). All these variations are good for the construction of number.[5]

"Duck, Duck, Goose" is not a game involving number as such. However, it contributes to the child's construction of number because it involves structuring many elements (the players) into an order. ("It" has to tap each child in a circle on the head and say, "Duck," until he picks a "Goose" to chase him or her around the circle.) The difficulty of this ordering can be seen among the four-year-olds who are generally less advanced cognitively. These children feel no necessity to touch each child in the circle, and skip one or more between those they dub "Duck." More advanced children, in contrast, are so careful not to skip anybody that they even feel impelled to go back to a child who wiggled out of place and got overlooked.

In "Fourteen Stand Still," each player takes 14 steps away from "It," who is blindfolded. After the fourteenth step, everybody has to stand still while the blindfolded "It" gropes to find someone. The teacher can encourage discussion to get the players to vary the number of steps to take. The smaller the number, the easier it is for "It" to find someone, and children decide whether they want to make this part of the game easy for "It" (and thus "dangerous" for themselves) or hard for "It" (and less "dangerous" for themselves).[6]

Hiding Games

The teacher can introduce addition and subtraction by asking children to guess how many out of 5 paper clips (or other small objects) he or she has covered up. The teacher begins by making sure children agree on how many objects there are, then asks them to close their eyes while some are covered. This is a good subtraction game not only because young children adore hiding games, but also because they can check their answer themselves by lifting the teacher's hand. This kind of feedback is far more convincing to children than a "C" that means "correct" and an "X" that means "incorrect." The ideal way to play this game is for children to set problems for one another.

A variation of this hiding game can be played with tongue depressors or ice cream sticks. In this variation, the teacher shows, for example, 7 sticks to the children, hides them under the table with both hands, and then brings out 4 in one hand. The children are asked, "How many are there in my other hand (or under the table)?" Once in a while, when the children guess correctly, the teacher should vary the game by putting one or two between his or her legs. If the children do not immediately suspect a trick when their anticipation is not verified, the teacher should playfully ask what happened. This is an ideal occasion for children to strengthen their notions of number.

(1) The children closed their eyes while the teacher hid none, 1, or more pennies in a game which was new to them.

(2) Three children had three different answers to the question: "How many do you think I have under my hand?"

"3!"

"1, 2, 3, 4!" (Pointing to each one of the pennies that were visible.)

"1!"

The teacher felt that this question was much too difficult for the children who gave the first two answers and decided to reduce the number to four the next time. (The first child's answer was a random guess. The second child's logic was not any better than the first one's.)

(3) To the two children who did not logically deduce the answer, the empirical proof made no more sense than finding out that soap floats sometimes and sinks sometimes.

(4) Being the "teacher" who hides enables children to think about number combinations from the perspective of the one who knows how many are hidden.

Rituals to Choose the First "It"

Rituals to choose the first "It" involve ordering and finding out who in this order corresponds to the last beat of the chant. A chant often used is the following:

> Eeney, meeney, miney, moe,
> Catch a tiger by its toe,
> If it hollers, let it go,
> Eeney, meeney, miney, moe.

When four-year-olds perform this ritual, their ordering is often completely random, just like the ordering illustrated in Figure 4(a). They frequently end the ritual by making themselves "It," no matter what the legitimate order should have been. Usually, it does not occur to the other players that there is something funny about this! This is another manifestation of the fact that four-year-olds do not feel any necessity of putting the elements into an order to make sure no element will be skipped and none will be counted more than once. This random pointing that makes oneself "It" is not due simply to the child's desire to be "It." Older children who feel obliged to follow a correct ordering try to figure out where to begin the ritual to make themselves "It." In the case of "Bubble Gum," which is described below, they figure out what number to say to become "It."

> Bubble gum, bubble gum,
> In a dish.
> How many pieces do you wish?

The child dubbed on the word "wish" must say a number. The children are then counted up to that number, starting with the next child in the order. To make himself or herself "It," the child who must say a number can say the number of children in the circle or any multiple thereof.

Art

In an occasional art activity, the children can each be given the same number of materials (such as 20 white beans, 20 red beans, 3 pieces of yarn, and some colored tissue paper), and encouraged to make a different collage. After the activity, the teacher can comment on how different children made such different designs with the same number of things.

Music and Movement

Songs and finger plays such as "Ten Little Indians" and "Five Little Monkeys" are excellent activities for teaching beginning number. Even before children understand the numerical meaning of the words, they have fun singing them in order as if they were arbitrary nonsense words. In "Five Little Monkeys," there is the added opportunity for counting backwards until none are left to jump on the bed.

So far, we have emphasized the incidental teaching of number in the context of activities which have other primary purposes. However, if not overdone, it is also a good idea to raise questions from time to time to get children to reason logically to get either an exact number or an estimate. Following are three examples.

> 1. When unpitted dates or cherries are available for snack, the teacher can ask casually, "How many cherries did you eat?" If children do not spontaneously count their pits, they can be asked, "How could we figure out how many each of us ate?" Candy wrappers might also be used in this way to figure out the exact number each person ate. For estimation, the teacher might put out quartered apples and encourage children to count each person's seeds to figure out how many of the pieces they probably ate. Also, children

can be asked, "How many seeds do you think there are in this apple (pumpkin, orange)?"

2. At group time or as children are waiting to be dismissed, the teacher might ask, "How many mittens (shoes, coats, or hats) do you think are in this room?" These questions are good because they encourage children to think about the different relationships between people and coats and between people and shoes. These questions can be complicated by taking into account the articles in the dress-up corner.

3. If a child is making a tower with blocks, the teacher might ask, "How many of these blocks do you need to make your height?" This is a question of number as well as measurement.

Notes

1. Cooking is not included among the desirable situations for the following reasons: The quantification that takes place in cooking usually involves continuous quantities (e.g., flour, milk, and sugar), which are beyond the scope of this paper. Also, children four to six years of age do not seem to care about quantifying these ingredients when they are cooking. For example, when they take a cup of water to make lemonade, many of them do not even notice when the cup is not full. For some, however, recipes do have educational value. We have seen a recipe written as follows:

2 ⊔ f

1 ⊔ s

1 ⊔ w

The first line says, "2 cups of flour," the second, "1 cup of salt," and the third, "1 cup of water."

2. Charts have the additional advantage of being good for the following purposes:

> Discussing why some children are absent and sending notes and pictures to them;
>
> Teaching how to read names;
>
> Teaching how to represent events. (That is, "present" can be represented with a "P" or " +, " and "absent" can be represented with an "A" or " −. " It is good for children to learn that there are many different ways of representing the same thing. Asking them to invent other ways of indicating "present" and "absent" would be good.)

3. We would like to mention another game we like but consider appropriate only for rather sophisticated children who have elementary number well in hand. In this game, "It" draws a card which has a number on it. The other children try to guess the number, and "It" responds with the feedback "more" or "less." For example, if the number drawn is five, and the first guess is "ten," "It" says "less." This game is inappropriate for unsophisticated children who would find it difficult to give or use the feedback. The problem with the game is that unless the child has a firm grasp of all the possibilities of "less than 10," he can only guess in random fashion. "It," too, may make errors in giving feedback, thereby confusing the guessers. If "It" makes an error in giving feedback, the other players have no way of judging the accuracy of "It's" judgment. Because this game is beyond the level of beginning elementary number, we omitted it from the text. In this paper, we tried to limit ourselves to activities involving objects and/or actions that are visible to all players.

4. The reader is referred to Piaget (1932, Chapter 1) for an elaboration of this developmental change.

5. We have tried other similar games in the form of races for objects where, for example, there is one block fewer than children. We do not recommend this type of race for very young children who have no

logico-mathematical notion of number. Even after being told why there will be one player who will not find a block, they seem to be surprised, puzzled, or crushed when they look for a block and find that there is one for everybody except themselves.

6. "Dangerous" may not be the appropriate term here because four- to six-year-olds like to be caught and want to be "It"!

In Conclusion . . .

In conclusion, we return to the question mentioned at the beginning of this book, "Isn't there any way the conservation task can be applied in the classroom?" Our reply is that this task helps the teacher to understand the logico-mathematical nature of number, and that if the child constructs the structure of number, he cannot avoid becoming a conserver. This structure cannot be "given" to the child, as the only way to have it is by constructing it from within. In this book, we tried to show that there are better ways to teach number than with kits, special materials, or workbooks, or by having a regular math time. We tried to give, instead, some examples of situations in which children are naturally motivated to quantify the stuff around them. The creative teacher will think of many other ways to pick up on this spontaneous interest.

References

Duncan, E. R.; Capps, L. R.; Dolciani, M. P.; Quast, W. G.; and Zweng, M. J. *Modern School Mathematics: Structure and Use.* Teacher's annotated ed., rev. ed. Boston: Houghton Mifflin, 1972.

Inhelder, B., and Piaget, J. *The Early Growth of Logic in the Child.* New York: Harper & Row, 1964. (First published as *La Genèse des Structures Logiques Élémentaires, Classification et Sériations.* Neuchâtel: Delachaux et Niestlé, 1959.)

Kamii, C. "Evaluation of Learning in Preschool Education: Socio-Emotional, Perceptual-Motor, Cognitive Development." In *Handbook on Formative and Summative Evaluation of Student Learning,* edited by B. Bloom, J. Hastings, G. Madaus. New York: McGraw-Hill, 1971.

Kamii, C., and DeVries, R. "Piaget for Early Education." In *The Preschool in Action,* edited by M. C. Day and R. K. Parker. 2nd ed. Boston: Allyn and Bacon, in press.

Kunz, J. *Modern Mathematics Made Meaningful with Cuisenaire Rods.* New Rochelle, N.Y.: Cuisenaire Co. of America, 1965.

Lavatelli, C. *Early Childhood Curriculum: A Piaget Program.* 2nd ed., Teacher's Guide. Boston: American Science and Engineering, 1973.

Montessori, M. *The Montessori Method.* New York: Schocken, 1964. (First published in English, New York: F. A. Stokes, 1912.)

Piaget, J. *The Child's Conception of Number.* New York: W. W. Norton, 1965. (First published with A. Szeminska as *La Genèse du Nombre chez l'Enfant.* Neuchâtel: Delachaux et Niestlé, 1941.)

Piaget, J. *The Moral Judgment of the Child.* New York: The Free Press, 1965. (First published as *Le Jugement Moral chez l'Enfant.* Paris: Alcan, 1932.)

Piaget, J. *To Understand Is to Invent* New York: Gross Publications, 1974.

Stern, C., and Stern, M. B. *Children Discover Arithmetic*. New York: Harper & Row, 1971.

Toys
"Candyland." Springfield, Mass.: Milton Bradley, 1955.
"Cat and Mouse." Salem, Mass.: Parker Brothers, 1964.
"Chutes and Ladders." Springfield, Mass.: Milton Bradley, 1956.
"Don't Spill the Beans." Minneapolis, Minn.: Schaper Manufacturing Co., 1967
"Hi-Ho! Cherry-O." Racine, Wis.: Western Publishing Co., 1972.

Notes